Gadgets Galore

The Latest High-Tech Inventions and Their Benefits

Table of Contents

Chapter 1. Introduction

In our Special Report titled "Gadgets Galore: The Latest High-Tech Inventions and Their Benefits," we delve into the vast and exciting world of cutting-edge technology. Carefully seated between the realms of imagination and reality, the latest gadgets have found ways to ace everyday tasks more efficiently and often, more joyfully. This report is an engaging journey, taking you through smart innovations and their transformative potential to enhance our routine lives. If you're fascinated by the future-forward tech potential and seeking to make informed decisions for your next gadget purchase, our special report will be your trusty guide. Ready to discover what the future holds? Welcome aboard! Let's navigate this exhilarating sea of high-tech inventions together. Expect intrigue, excitement, and above all, knowledge that's worth every penny!

Chapter 2. Meet the Giants: A Look into Leading Tech Manufacturers of Today

The cutting-edge technology that we enjoy today is majorly the brainchild of some towering tech manufacturers who are at the helm of these remarkable advancements. These leading players constantly push the boundaries of what is possible, thriving on a spirit of constant innovation, creativity, and uncompromising quality.

At this juncture, let's unravel the stories behind these tech giants, exploring their innovations, ethos, and the future outlook they convey.

2.1. Apple Inc.

Recognizable by its iconic bitten-apple logo, Apple Inc., the American multinational technology company, has always focused on innovation and sleek aesthetics. Its story began in the garage of co-founder Steve Jobs in the 1970s.

Apple's iPhones, Macs, iPads, and Apple Watches are loved for their design elegance, ease-of-use, and performance. Apart from hardware, Apple's software, and online services, which include iOS, iTunes, and iCloud, have also profoundly impacted the tech-scape. With continuous innovation, Apple promises to unveil amazing user experiences with unique tech integrations in the future.

2.2. Samsung Electronics

Hailing from South Korea, Samsung Electronics has been a formidable player in the electronics industry since 1969. With a wide array of products ranging from smartphones, televisions, and refrigerators, to semiconductors, it has left an indelible mark in the pages of tech history.

Samsung's Galaxy series has emerged as a significant competitor to Apple's iPhone. Besides, Samsung's advancements across customers and industrial electronics, including the trending Smart Home solutions, have been commendable. Being the world's largest memory chipmaker adds another feather to the Samsung cap.

2.3. Microsoft Corporation

Microsoft Corporation needs no introduction for anyone with minimal exposure to computers. Founded by Bill Gates and Paul Allen in 1975, Microsoft morphed into a tech juggernaut, shaping the way we understand and experience personal computing.

Microsoft Windows - arguably, its flagship product - remains the backbone of most PCs worldwide. Office Suite, Xbox gaming range, Surface devices, and cloud services Azure - each extends Microsoft's sphere of influence. Productivity, AI, and cloud are among Microsoft's keen focus areas for future advancements.

2.4. Alphabet Inc.

What began as a search engine named "Google" by Larry Page and Sergey Brin in 1998, transformed into a conglomerate, Alphabet Inc. Its myriad entities influence advertising, cloud computing, artificial intelligence, and beyond, but Google, with its search, Maps, Android, YouTube, and Google Cloud, remains its star.

Google's upcoming projects like AI-driven solutions and quantum computing speak volumes about its futuristic vision. With advancements that permeate through almost all digital aspects of human life, Alphabet continues to redefine how humans and technology interact.

2.5. Huawei Technologies

As China's largest technology company, Huawei Technologies has evolved significantly since its inception in 1987. Initially a phone switch producer, Huawei has since extended its reach to telecommunications networks, devices, and cloud computing.

Its smartphone range- primarily the Mate and P series- competes well in global markets. However, it's Huawei's robust work in 5G technology that has catapulted its image as an industry pioneer. Amid all odds, Huawei's commitment to a fully connected, intelligent world remains steadfast.

2.6. Sony Corporation

From Walkman to PlayStation, Sony Corporation has been ruling hearts with its innovative gadgets for decades. Established in Japan in 1946, Sony's commitment to "kando" - emotional involvement - has been the secret ingredient to its iconic technological wonders.

Sony's technological prowess extends over numerous electronics, including TVs, audio equipment, imaging products, gaming consoles, and more. Innovation, as Sony believes, is endless. And so, we may look forward to even more trailblazing advancements from them in the coming years.

These giants have proven that they are the bedrock of the current technological revolution, unveiling unprecedented gadgets that shape our lives. With their fingers firmly on the pulse of innovation, one can only imagine the technological marvels we're set to encounter in the future. However, these manufacturing bigwigs are far from the only players. Newer companies bring fresh perspectives and groundbreaking inventions to the mix, keeping the tech game fascinatingly unpredictable.

Witnessing the journey of these tech titans unravels their contribution in pushing technology towards unexplored horizons. The future sure looks exciting, with the wheel of innovation constantly in motion, propelled by these titans and others who dare to dream and innovate.

Chapter 3. The Smart Home Revolution: Transforming Living Spaces

From commanding Alexa to play your favorite tunes, to asking Siri about tomorrow's weather, smart technologies have gradually integrated into our homes, seamlessly merging into our everyday lives. Possessing a blend of convenience, security, and fun, they have redefined what it means to be 'home', evolving it into a space that learns, adapts, and engages intuitively.

3.1. Enter the Age of Smart Homes

The beginning of the smart home era could be traced back to 1966 with a system called "Echo IV". It was humankind's first glimpse into automated living. From then onwards, the journey has been nothing short of revolutionary. In the 1970s, technologies such as X10 allowed devices to communicate over electrical wiring. Fast forward to now, we've a myriad of wireless technologies allowing for greater flexibility and broader applications.

Supreme among the wireless technologies is IoT (Internet of Things). Sensing, processing, and communicating data, IoT devices make homes responsive and adaptive. Think of your home as a symphony with multiple IoT devices as musicians collaborating to orchestrate your life.

3.2. IoT: The Conductor of a Smart Home Symphony

Each IoT device collects data about their specialized area of operation, and together they not only optimize the individual operations but also harmoniously integrate for overall home optimization. Your thermostat understands the weather and continually sets home's temperature, your refrigerator knows your consumption patterns and orders groceries accordingly while your safety systems monitor for anomalies and alert you in real time to keep you secure.

The essence of IoT lies not just in automation but in communication. Devices "talk" to one another to ensure a seamless interaction. This is where hubs or smart speakers like Alexa and Google Home act as a universal translator coordinating between devices.

3.3. The Magic of Smart Assistants and Hubs

Smart assistants have an integral role in a smart home infrastructure. They can connect to various smart devices, but also provide an interface for human interaction — voice. Voice technology not only offers a hands-free experience but also personalization based on voice recognition. Need your favorite playlist to start your day? Or want to set up a unique ambiance for your dinner date? Just command your smart assistant!

Additionally, these hub devices often come with mobile applications which provide a visual overview and remote control of your devices. This ensures you are connected to your home, no matter where you are.

3.4. The Different Facets of Smart Homes

Smart homes have monumental potential spanning across various aspects of human lives.

Energy efficiency takes center-stage with smart thermostats adjusting temperature according to routines and weather, smart lights doing the same with ambient lighting, and devices optimally utilizing power resources using sensors and AI. All aimed to reduce carbon footprint and save on energy bills.

Safety and security devices, like smart locks, security cameras, smoke detectors, and other sensor-based devices, provide peace of mind. They not only monitor for unwanted intrusion but also for any potential hazards lurking inside.

Home healthcare is another emerging application. Wearable devices and health monitors can keep a real-time tab on health vitals of residents. They can alert for emergencies and also share data with healthcare providers for regular checkups or consultation.

Personalized comfort and convenience is a given. Why stick with one-size-fits-all when you can customize everything according to your needs? Be it lighting, music, temperature, or even your coffee machine.

3.5. Delving into the Mind: Artificial Intelligence in Smart Homes

As we captivate this thriving landscape of smart homes, it becomes incumbent to look under the hood and perceive the engine that powers it, i.e., Artificial Intelligence (AI). Colossal in its implications, AI serves as the brain of a smart home ecosystem. While the IoT provides the sensory input, AI enables the home to 'think' and 'learn'.

AI enables pattern recognition, predictive analysis, and machine learning. This means your Smart Home system can learn from your behavior and preferences, create predictive and reactive automation, and provide data-driven insights about your lifestyle. The experience thus becomes more effortless, efficient, and personalized over time.

3.6. Looking Ahead: The Future of Smart Homes

The future of smart homes is luminous. As AI and IoT technology progress, we can only imagine the possibilities. Ever more devices will become 'smart,' the intercommunication will become more seamless, and AI will become more refined in adapting and predicting.

Interoperability could unlock cross-application interactions taking integration to new heights. For example, your sleep tracker could communicate with your coffee machine to get your morning blend ready just as you wake up. Or your intelligent kitchen could suggest recipes based on the ingredients in your smart fridge.

While we need to thread with caution navigating through concerns of cyber-security and data privacy, the potential reward safeguards the pursuit. For all we know, we might merely be at the brink of 'intelligent living.'

Conclusively, the Smart Home Revolution is not just transforming our living spaces; it's transcending them. It's augmenting our homes to be an extension of our personal selves, catering to our comforts, minding our safety, and enhancing our sustainability. It's leaving no stone unturned in its mission to bewitch us with the perfect blend of technology, utility, and innovation. The future, it appears, is undoubtedly promising, undoubtedly bright, and undoubtedly, smarter. The voyage forward to the next frontiers of smart living promises to be as riveting as it is remarkable. So, as the smart home landscape continues to burgeon, the roles we play around it, with it, and within it, continue to exponentially evolve. Welcome to the new era of living!

Chapter 4. Wearable Tech: More Than Just Fitness Trackers

There's much more to wearable technology than fitness trackers and smartwatches. While most of us think of these devices when we hear "wearable tech," innovators are going beyond the realms of step counting and heart rate monitoring.

4.1. Evolving Landscapes

Formerly a sector defined by fitness and wellness devices, wearable technology is finally stepping out of its shell, diverse applications have eased into areas such as medical diagnostics, elderly care, virtual reality, fashion, and more. This new wave of wearable innovations blends technology with our daily lives, going beyond mere convenience to offer critical real-time data and proactive healthcare supports.

4.2. Wearable Tech in Healthcare

Wearable technology plays a vital role in modern healthcare. For instance, 'EKG monitors' make cardiac care accessible at home by allowing users to capture a medical-grade EKG in just 30 seconds. Blood pressure monitors embedded in wrist watches predict irregularities and keep track of heart health in real time. Electrodermal activity sensors are useful in monitoring mental stress levels.

For those living with diabetes, 'smart' contact lenses could make glucose level monitoring a breeze. The technology measures glucose levels in tears and alerts users when they need insulin. Also, there are smart patches, which release medication in response to changes in skin temperature or sweat analysis, a potential game-changer for people living with chronic conditions.

4.3. Senior Care & Wearable Tech

The wearables market also cares for the elderly and aging population. With devices capable of detecting falls and signaling for help, wearables provide peace of mind to seniors living independently. GPS-enabled wearables are useful for tracking patients with Alzheimer's or dementia, who may wander off or forget their way home. Some even come with built-in medical alert systems, further improving elderly care.

4.4. Virtual Reality Wearables

Technology is reshaping our perception of reality. Virtual reality (VR) and augmented reality (AR) wearables are becoming increasingly prevalent and have far-reaching implications for entertainment and gaming. VR headsets transport you to entirely different worlds for immersive experiences, while Mixed Reality (MR) devices blurs the line between the virtual and real world.

However, these devices aren't just for play; learning and training applications are equally impressive. From medical students practicing complex surgeries in a controlled virtual environment to soldiers carrying out military simulations, VR is changing traditional learning methodologies forever.

4.5. Fashionable Tech

Fashion and technology are not mutually exclusive, as proven by fashion-forward wearables. For instance, designers have cultivated 'smart' jewelry that serves as a panic button or tracks UV radiation levels. Clothing equipped with NFC (Near Field Communication) allows wearers to share information digitally, enhancing networking at social gatherings. More intriguing are color-changing fabrics, responding to changes in mood or environment, likely to redefine fashion standards in the future.

4.6. Future of Wearable Tech

With rapid advancements and broader acceptance, the future of wearable tech is promising. We might see more advanced features integrated into smartwatches, alongside the development of application-specific wearable gadgets. Devices might get smaller, more discreet while offering improved functionality.

We can foresee a future where wearable technology will emancipate us from screens, providing an interactive digital experience throughout our day-to-day lives. Moreover, these devices will deliver superior healthcare, remote patient monitoring, better data-driven treatments, and ultimately, improve the quality of life and longevity.

So, are you ready to upgrade your lifestyle with wearable tech? With the innovation wheel turning at a breakneck speed, there's no better time than now to explore and adopt the renaissance of wearable tech. The future, it seems, is already strapped to our wrists, clipped on our clothes, and wrapped around our reality - and it's possibly only the beginning.

Chapter 5. Virtual Reality & Augmented Reality: Beyond Games and Entertainment

The frontier of Virtual Reality (VR) and Augmented Reality (AR) has expanded beyond what we imagined a decade ago. Initially, most associated these technologies with the gaming and entertainment industries. However, the vivid horizons of VR and AR now extend far beyond these domains, offering remarkable opportunities in various sectors, ranging from healthcare, education, and manufacturing to architecture, real estate, and tourism.

5.1. The Principles of VR and AR

Before venturing into the myriad applications of these two technologies, let's grasp the underlying principles. VR displaces us from the actual world and immerses us into a simulated environment. When in a VR state, users are isolated from the real world with the use of VR headsets. Sight and sound are the primary senses used, although more advanced systems provide tactile feedback for complete immersion.

On the other hand, AR integrates computer-generated enhancements into the real world to make it more interactive and digitally controllable. This combination of real-world and virtual elements is achieved through the use of smartphones, smart glasses, and AR headsets.

5.2. Applications in Healthcare

VR and AR have found substantial applications in healthcare, ranging from treatment to training. Medical practitioners use VR to simulate surgeries for practice and to plan real ones, leading to reduced operation times and improved surgical outcomes. AR can aid surgeons during procedures by providing real-time 3D images of the patient's anatomy, eliminating the need for invasive measures to determine the best surgical approach.

Psychologists and therapists are also increasingly using VR to treat various disorders, such as Post Traumatic Stress Disorder (PTSD) and phobias, by generating controlled environments for exposure therapy.

AR is a boon for medical education too; structures and functions of complicated anatomical structures can be explained more effectively using 3D models. Students can witness intricate surgical procedures in real-time without actually being present in the operation theaters.

5.3. Use in Education

The use of VR and AR is steadily increasing in the arena of education. Both technologies offer immersive learning experiences, making understanding of complex concepts and subjects more enjoyable and less taxing for students. With VR, physics can be understood by virtually visiting the Large Hadron Collider, or history can be learned by walking through a digitally replicated Ancient Rome.

AR also enhances classrooms by adding interactive layers to traditional methods. Children can interact with 3D models of objects, animals, or even celestial bodies, reducing the gap between theoretical knowledge and practical understanding.

5.4. Scope in Manufacturing and Prototyping

In fields where production costs are high and precision is key, VR and AR prove to be game-changers. Companies can design and test their products using VR before starting actual production, thereby reducing the cost of making physical prototypes.

AR can overlay virtual prototypes onto the real world environment, testing size, fit, and design before actual production begins and thus mitigating the risk of design failure.

5.5. Invasion into Architecture and Real Estate

Before actual construction, architects can now present their designs in 3D using VR. Prospective clients can virtually tour the property, helping them to accurately envision the completed project.

In real estate, prospective home buyers can virtually and remotely tour multiple properties before making a decision, saving travel time and making house hunting more convenient.

5.6. Advancements in Tourism

Travel companies and tourism boards are leveraging VR to offer previews of destinations to potential tourists. This immersive sneak peek helps travelers make an informed choice about their travel destinations. Museums and historical sites are also using AR to enrich the visitor experience by overlaying tour information to the real world, thus making tours more interactive and informative.

In the nutshell, VR and AR technologies have moved beyond the realm of games and entertainment, revolutionizing numerous sectors with their innovative capabilities. As technology continues to evolve, it's safe to say that this is just the beginning. There's a vast sea of unlimited potential awaiting us in the fields of VR and AR, and the journey of exploration has just set sail. Buckle up for a ride into the exciting future!

Chapter 6. Drones and Robotics: Automating the Future

In our world today, drones and robotics occupy the center stage of the technological theatre, driving the narrative towards a future where automated systems become ubiquitous in our daily lives. These are the marvels of engineering and artificial intelligence, beautifully merging to craft intricate solutions for tasks ranging from the mundane to the complex.

6.1. Mechanisms of Drones and Robotics

At their very core, drones and robotics operate on similar principles. They are designed to be autonomous or semi-autonomous, with an ability to perform tasks without human intervention. The components creating these spectacular pieces of technology are fascinating.

Robots have been around for decades, and their evolution is a testament to human ingenuity. The fundamental elements of robotics are actuator (which is analogous to muscles), sensor (which is similar to human sense organs), controller (akin to the brain), and power supply (comparable to the food we eat). The interaction between these components makes a robot behave in the way it does.

Aerial drones, also called unmanned aerial vehicles (UAVs), are operated either remotely or autonomously. Their anatomy comprises the flight controller functioning as the brain, sensors such as GPS and gyroscopes aiding navigation, battery for power, propellers for mobility and a shell that holds it all together.

6.2. Etching Footprints on Varied Fields

The scope and benefits of drones and robots are wide-reaching, touching various sectors of our society, proving the technology's versatility.

In agriculture, drones have transformed the way we farm. Detailed mapping, crop surveillance, and precision spraying are just a few of the tasks drones have made more efficient. Autonomous tractors can work all day without getting tired, ensuring productivity levels unheard in traditional farming.

In healthcare, robots have made significant strides. From performing microsurgeries with astounding precision to assisting in patient care, the influence is evident and growing. Robotic exoskeletons help patients with mobility issues regain independence, while telepresence robots enable healthcare professionals to virtually interact with patients in remote areas.

In military applications, drones and robotics command an established presence. Reconnaissance missions, precision strikes, demining missions, and surveillance are a few of the critical functions performed by robotic technologies in defense.

The list of applications hardly ends here. In search and rescue missions, logistics, construction, entertainment, and more, it is clear that drones and robotics are indeed automating the future.

6.3. The Ethical Scene Surrounding Automation

The increasing reliance on drones and robotics does come with pressing ethical concerns. Appreciating the transformative potential of these technologies, it is crucial to approach their deployment with empathy, caution, and care.

Job displacement is a significant concern. As robots become capable of performing complex tasks, the fear of unemployment for certain job categories is real and worrisome. It's crucial to strike a balance and ensure that the benefits of technology reach everyone, without fostering socioeconomic disparity.

Privacy issues related to drone usage are another ethical controversy. The ability of drones to reach potentially any corner raises concerns over surveillance and personal privacy invasion. Regulations on drone usage are needed to mitigate this.

The discourse on autonomous machines in warfare also raises questions about accountability. In a scenario where an autonomous machine causes harm, identifying culpability becomes difficult.

6.4. Emergence of Drone and Robot Economies

Despite the challenges, the potential for drones and robotics to create new economic industries is immense. From service industries centered around robot maintenance, software upgrades, and part replacements, to transforming existing industries like delivery and logistics, there's a tremendous global economic potential.

Moreover, the increase in productivity provided by drones and robotics has the potential to revolutionize the global economy, with an estimated raise of up to 1.5% GDP per year.

Revenue influx from drone-related activities, like drone photography, mapping, and inspection services, have created new avenues for entrepreneurs and businesses. Similarly, in the robotics sector, revenues from robot sales, automation services, and related activities are injecting life into what's being hailed as the 'Robot Economy'.

6.5. The Road Ahead

The convergence of technologies like artificial intelligence, big data, and sensor technology is likely to take drones and robotics to new heights. The very notion of drones and robots is changing as these systems become more autonomous and intelligent.

Innovation in energy technologies can lead to more efficient fuel cells or solar-powered drones. Similarly, advancements in AI could lead to more intelligent robots capable of learning from experience.

The future holds immense promise. As we stand on the brink of this exciting era, it's essential to embrace these technologies while acknowledging and addressing their challenges. The balance between efficiency and ethics, between convenience and privacy, will determine how drones and robotics continue to shape our collective future.

Irrespective of any hurdles, it is evident that drones and robotics are here to stay, automating us into a future fraught with opportunities, innovation, and progress. The convergence of man and machine, if navigated adroitly, can orchestrate a symphony of growth, prosperity, and human welfare. This automation drive, therefore, warrants both our applause and our astute attention.

Chapter 7. Green Gadgets: Eco-friendly Technology and Sustainability

In an era where environmental awareness is no longer an option but a responsibility, the technology landscape is awash with innovative, eco-friendly gadgets. These contrivances not only exhibit cutting-edge technology but also create a minimal ecological footprint, making them the prime choice for a sustainable future.

7.1. Solar Power Banks

A shining example of green tech innovation is solar power banks. Now, you may never have to worry about your device battery dying when you are away from a plug, as long as you have sunlight. The solar power banks employ photovoltaic panels to convert sunlight into electricity, which is then stored for future use, thus making them a renewable energy-based charging solution. From powering smartphones, tablets to even laptops, this product addresses the off-grid power needs while greening our tech-intensive lives.

7.2. Smart Power Strips

Regular power strips continue to consume electricity even when your devices are fully charged or on standby. Smart Power Strips, on the other hand, eliminate this wastage by auto-detecting and cutting off power supply when the devices are not in use or fully charged. Some models can even differentiate between many devices and can provide dedicated energy-saving strategies for each.

7.3. Eco-friendly Printers

Eco-friendly printers are a remarkable replacement for the conventional energy-guzzling printers. They use sustainable techniques such as refillable ink cartridges and duplex printing, thereby minimizing paper waste. Certain models have a standby mode when not in use, which reduces energy consumption. Some are designed to use soy-based inks, a renewable resource, thus lowering their carbon footprint.

7.4. Energy-saving Light Bulbs

Moving from incandescent bulbs to energy-saving light bulbs, like LEDs, is one of the simplest ways to make your home greener. LEDs consume up to 80% less electricity compared to traditional bulbs and have a considerably longer lifespan, often lasting for over a decade depending on usage. They even come in a variety of color temperatures, allowing for customization based on personal preferences or ambience requirements.

7.5. Green Chargers

Breathing new life into the concept of chargers are the innovative green chargers. These range from hand crank models, which generate electricity through simple mechanical effort, to wireless electromagnetic induction chargers, which reduce electronic waste by not requiring different cables for each device. Whichever type you choose, green chargers deliver sustainability alongside functionality.

7.6. Home Energy Monitors

For those looking to better understand and manage their household energy consumption, home energy monitors serve as a great tool. These smart devices provide real-time data about your home's energy usage, making users aware of the energy vampires in their households and assisting in optimizing energy consumption. This data can be viewed conveniently on a compatible smartphone or tablet.

7.7. E-Readers

E-readers have already proved to be a game-changer for voracious readers. The ability to store thousands of books digitally not only considerably cuts down paper usage but also reduces the carbon emissions associated with shipping physical books. With e-ink technology that mimics the readability of actual paper, these devices provide a compelling blend of eco-friendliness and convenience.

7.8. Wind Up Radios

Dynamo-powered wind-up radios free users from the dependency on electricity or batteries. With just a few turns of the crank, these radios can run for several hours, making them not only a sustainable alternative to battery-powered radios but also a crucial tool for emergency preparedness.

In a digital economy where technology and gadgets form the bedrock of our daily lives, the ramifications of our tech choices on the environment are more significant than ever. While the gadgets discussed in this report represent a promising start towards marrying innovation with sustainability, the future surely holds even more groundbreaking, eco-conscious technology. However, ultimately, the quest for a greener tomorrow heavily relies on our responsible adoption and usage of these technological gifts. Hence, let's plug into this green tech revolution with an open mind and a commitment to reducing our ecological footprint!

Chapter 8. Artificial Intelligence: Gadgets that Learn and Adapt

In the realm of technology, few domains spark as much curiosity and excitement as Artificial Intelligence (AI). Far-reaching in its scope, AI has pervaded several aspects of modern life, bringing about profound changes in everything from business to our daily routines. In living rooms, workplaces, or on the go, AI-enabled gadgets are becoming increasingly popular, each equipped with the ability to learn from user input and adapt accordingly. They subtly improve task handling and decision making, laying a robust pathway towards a technologically intervened future.

8.1. Understanding AI: More than Just a Buzzword

Before delving into the AI-enabled gadgets themselves, let's build a base understanding of what constitutes Artificial Intelligence. AI refers to computer systems capable of performing tasks generally requiring human intelligence, including perceiving their environment, understanding natural language, problem-solving, and learning from experience.

AI can be categorized broadly into two types: narrow AI, which excel in specific tasks (e.g., recommending songs on music apps or managing your smart home), and general AI, which can theoretically perform any intellectual task a human can do, a realm we've yet to fully explore.

8.2. AI-Enabled Gadgets: A New Wave of Smarter Devices

AI's presence has increasingly become more pronounced and ubiquitous through the implementation within various everyday gadgets.

Voice assistants like Amazon's Alexa, Google Assistant, or Apple's Siri have become household staples. These AI-based tools go beyond just executing commands from the user; they learn from past interactions to furnish better, more useful responses each time. Whether it's personalizing music choices, suggesting recipes, or adjusting smart home settings based on individual preferences, these gadgets continually adapt to serve you better.

AI takes the guesswork out for fitness enthusiasts too. Products like Fitbit and Apple Watch have AI capabilities that analyze vast amounts of data, from tracking heart rates to monitoring sleep patterns, helping users understand their health patterns and adapt their lifestyles for optimum wellness.

In the gaming arena, new-age AI-powered consoles bring immersive, adaptive experiences like never before. Google's DeepMind, for example, uses AI that learns from player behaviors and adapts to provide an ever-challenging, exciting gaming experience.

Artificially intelligent cameras offer smart surveillance and photography capabilities. For example, Google's Nest cameras set benchmarks for AI integration into home security, learning to distinguish between different types of motion and alerting homeowners accurately.

8.3. The Seamless Learning Curve: How AI Gadgets Learn

A key characteristic of AI gadgets is their adaptability. Unlike traditional devices, these gadgets can 'learn' and 'grow.' Each interaction provides data that the system analyzes to improve future functioning. This learning process, often called 'machine learning,' lies at the core of AI tech.

It starts with data collection. Once an AI gadget is operational, it begins gathering data from each user interaction. This could be anything: a prompt to play a specific song on a smart speaker, changes in heart rate monitored by a fitness tracker, or even your playing style in a video game.

The system uses this data to compute patterns and correlations. Advanced algorithms process this collected data to discern actionable patterns. Over time, the AI system learns to predict your behavior, thereby adapting its functioning to meet your needs more effectively.

8.4. The Ethical Elephant in the Room: Data Privacy and AI

As AI tech moves into uncharted territories, concerns around data privacy and ethical use of AI have moved to the forefront. As AI gadgets learn and adapt, they collect and analyze vast amounts of personal data from the interactions and inputs given by the users.

While regulations like Europe's General Data Protection Regulation (GDPR) force companies to be transparent about their AI data use, it is crucial as an informed user to understand and control the data shared with these gadgets.

8.5. Ensuring AI is Actually Intelligent: Gadgets that Pass the Test

A gadget claiming AI capability does not always mean it possesses advanced, adaptive learning technology. Many companies, in recent years, have been criticized for 'AI washing'—the overuse or misuse of the term 'AI' to appear technologically advanced.

Hence, before investing in AI gadgets, research their abilities. Look for devices that offer machine learning capabilities, user adaptability, and, importantly, robust data privacy protocols. Transparent communication from manufacturers about how and what data is collected, processed, and stored is a key indicator of a device's trustworthiness.

8.6. AI: The Way Forward

AI's potential to revolutionize our lives seems endless as it continues to seep into various aspects of our day-to-day. Artificial Intelligence has moved beyond being a theoretical concept to become a tangible reality, offering advanced solutions to simple problems and simplifying complex tasks.

In the next few years, we await a universe where AI technology is not special but standard—a world where our gadgets do more than just function; they understand, adapt, and evolve with us. The future of AI is here, and its promising vista is only just over the horizon, mandating our active engagement with it and its exciting possibilities.

Remember, informed decision-making today will result in a better, technologically harmonious tomorrow. So, as you examine the endless ocean of possibilities that technology has to offer, may this report shine as a beacon, guiding you towards your perfect gadget match.

Chapter 9. Health-Focused Innovations: Gadgets for a Healthier Tomorrow

In a world increasingly conscious of health and wellness, there's an ever-expanding list of gadgets that have found fascinating ways of tapping into this trend. From wearables focused on activity tracking to devices assisting in mental well-being, the landscape of health-focused technology is vast, diverse, and ever-growing.

9.1. Fitness Trackers and Smartwatches

In the world of wearable technology, fitness trackers and smartwatches hold a prominent position. Companies like Fitbit, Garmin, and Apple have been releasing increasingly sophisticated devices designed to not just track activity, but also provide a host of other health-related information.

Apple's Watch Series 6 is revolutionary in its own way. Integrated with the most deeply personal health and wellness tools, it features an oximeter to measure oxygen saturation in your blood, an ECG app, quality sleep tracking, and fitness programs catering to various requirements. Furthermore, it offers seamless interaction with your iPhone and Apple's ecosystem, acting as a powerful addition to the user's interconnected tech life.

Garmin's wide range of fitness trackers and smartwatches have a more sport-specific focus. Devices like the Fenix 6 Pro offer advanced training metrics, including running dynamics, heat and altitude-adjusting VO2 max, and a recovery advisor. They're a great choice for those focused on particular sports or training for specific events.

Fitbit's Charge 4 provides robust health and fitness features, including built-in GPS, sleep tracking, heart rate monitoring, and compatibility with both Android and iOS. With Spotify integration and up to 7 days of battery life, it's a worthy investment for those seeking to optimize their health and wellness.

9.2. Health Monitoring Devices

Beyond fitness tracking, several devices are designed to monitor health conditions and provide actionable insights.

One such device is the Withings Body Cardio, a smart scale that measures not only weight but also body composition, heart rate, and even the weather. It offers an insight into your cardiovascular health through pulse wave velocity, which can help in early detection of health issues.

Another commendable device is the QardioArm blood pressure monitor, which pairs with your smartphone and records all blood pressure data, making it easy to share with your doctor. It's compact, portable, and makes blood pressure management hassle-free.

For diabetes management, the FreeStyle Libre is a game-changing gadget. It's a glucose monitoring system that doesn't require traditional finger-prick tests, instead using a small sensor worn under the skin to read glucose levels.

9.3. Mental Well-being Gadgets

In our stressful, fast-paced modern lives, mental well-being is as crucial as physical health. Unsurprisingly, numerous gadgets have been launched to assist with stress management, mental clarity, and improved sleep.

Headspace and Calm, leading meditation and sleep apps, offer guided meditation programs, sleep stories, breathing techniques, and more to promote a healthier, more relaxed state of mind.

For a more hands-on approach, the Muse 2 headband is a unique gadget. It uses advanced EEG technology to provide real-time feedback on your meditation practice, helping you enhance your focus. Coupled with guided meditations and soothing soundscapes, it's a remarkable tool for attaining mental tranquility.

The URGONight headband is another noteworthy gadget that aims at improving sleep. It trains your brain to produce brainwaves associated with natural, restful sleep, resulting in improved sleep over time.

9.4. Nutrition and Hydration

Understanding and managing diet is a key aspect of health and wellness. With modern technology, it's becoming easier to monitor and improve our nutrition and hydration.

The Smart Nutrition Bottle by LifeFuels enables users to create personalized beverage blends using flavor and nutrient Pods. It syncs with fitness wearables, recommending hydration levels based on activity.

The HAPIfork is an electronic fork that helps pace your eating. It vibrates and flashes lights if you are eating too fast, helping you gain control over your eating habits and promote weight loss.

9.5. Final Thoughts

From wearable tech to dedicated health monitoring devices, mental well-being assistance to nutrition and hydration, health-focused innovations are playing an increasingly integral role in our lives. As technology continues to advance rapidly, this trend is expected only to strengthen. Navigating through the extensive world of health gadgetry might seem daunting, but with informed decisions, these gadgets can lead to a healthier, better-informed, and more efficient lifestyle.

Chapter 10. Automobile Technology: Driverless Cars and Beyond

Modern society's journey on the road of transportation has been a wild one, going from humble horse-drawn carriages to futuristic visions of flying cars. Yet nothing has quite captured the imagination of tech enthusiasts, environmentalists, logistics companies, car manufacturers and governments like driverless cars have.

10.1. Understanding Driverless Cars

Before jumping directly into the advanced realm of automobile technology, let's start with basics: 'What is a driverless car?'. And, for that, we need to know a little bit about the vehicle automation levels.

Driverless cars, also known as autonomous cars or self-driving cars, combine sensors and software to control, navigate, and drive the vehicle independently. According to the Society of Automotive Engineers (SAE), there are six levels of driving automation, from 0 (fully human control) to 5 (fully automated system). At Level 0, there's no automation at all, the human does everything. Level 1 and 2 include advanced driver-assistance systems (ADAS), where vehicle safety features, like adaptive cruise control and self-parking, are automated, yet human intervention is required when these systems request. Level 3 to 5, however, are the zones where autonomy truly begins to shine.

Level 3 automation can handle all driving tasks, yet human intervention may be required in certain circumstances. Level 4 automation, on the other hand, can perform all tasks that a human driver can execute under certain conditions. Level 5 is the magic number - it's fully autonomous and requires no human attention, meaning the car can do everything a human driver can, regardless of the circumstances or environment.

10.2. The Sensing Technology

One of the largest hurdles for autonomous vehicles is getting them to recognize and react appropriately to their surroundings. After all, the world can be an unpredictable place, full of cyclists, pedestrians, other vehicles, and a myriad of other obstacles.

To understand and interact appropriately with the environment, autonomous cars rely heavily on advanced sensing technologies. LiDAR (Light Detection and Ranging), Radar (Radio Detection and Ranging), Cameras, Ultrasonic sensors, and GNSS (Global Navigation Satellite System) are some of the vital technologies used to detect obstacles, read road signs, distinguish between different objects, and locate the vehicle's exact position on Earth.

LiDAR works by sending out a million laser points per second and measures the time it takes for the signals to bounce back, calculating distance by utilizing the speed of light principle. Radars, on the other hand, use radio wave frequency to detect the location of obstacles. Cameras visualize lane marking, street signs, and traffic signals while ultrasonic sensors help detect objects in close range. Lastly, GNSS, similar to GPS but more accurate, helps maintain the vehicle's positioning and navigation.

10.3. How Autonomous Cars Work

The true magic lies in uniting these distinct technologies to work in harmony, making the car truly autonomous. The software that self-driving vehicles use is called an automated driving system (ADS). In short, this software processes the data from different sensors, makes decisions, and controls the car accordingly.

Imagine it like a cycle. Firstly, the perception step happens where LiDAR, radar, cameras, and ultrasonic sensors collect data about the car's immediate environment. This information is then fused with GNSS data, which maps the car's location precisely on the planet. Then, in the localization step, the vehicle determines its exact position relative to its surrounding environment using maps and sensor data. This is followed up by the planning stage, where algorithms decide the vehicle's path, considering things like lane changes, turn-taking, speed limits, and following rules of the road.

Next, the control function happens where the ADS software sends control commands to the vehicle's actuators controlling steering, braking, and acceleration. Finally, the feedback step ensures that the executed action aligns with the one planned. If it doesn't, the system changes the driving strategy, and the cycle repeats.

10.4. The Benefits and Challenges

Of course, this all sounds incredibly futuristic. However, there are both potential benefits and challenges that need to be considered when discussing autonomous cars.

Benefits include increased safety by reducing human error, which is a leading cause of accidents. The elderly or individuals with disabilities, who might not otherwise be able to drive, will have newfound freedom. Not to mention, productivity could increase given that time spent driving could be spent doing other tasks.

Conversely, challenges range from high costs and job displacements to security and regulations. Replacing human drivers with autonomous vehicles could threaten millions of jobs. There are also legal and hypothetical questions about an autonomous vehicle's decision-making, for instance, how an autonomous car should respond in unavoidable accident scenarios.

10.5. The Road Ahead

While we are still a few years away from widespread, day-to-day operation of autonomous vehicles, progress, and optimism are sky high. With tech giants and automakers spending billions, autonomous cars are more a question of 'when' rather than 'if'. As we drive down the road of the future, there's clear reason to remain excited about the prospects of these vehicles.

In conclusion, driverless cars are a marvel of modern technology, perfectly embodying the convergence of hardware and software to create devices that can potentially revolutionize our method of travel, promising safer roads and more efficient journeys. They are part of an exciting technological future where tasks once totally human-controlled are being passed over to automated systems, and the reality of this high-tech invention is no longer a distant dream, but increasingly an imminent reality. The extensive landscape of autonomous vehicle technology is vast and intricately woven. Today's exploration provided a comprehensive view, but like the very technology we covered, our knowledge continues to expand, advance, and evolve – and there is much more to explore and master.

Let the automobile technology revolution begin, and may we meet again on the roads of tomorrow!

Chapter 11. Future Outlook: Emerging Tech Trends and Their Potential Impact

The revolution of technology is a tale told by humanity over the centuries. Emerging technologies are rapidly evolving, and the accelerating pace of change presents novel opportunities and threats - economically, socially, and of course, technologically.

11.1. The Dawn of the Quantum Age

A new era of technology is set to redefine our understanding of computation itself - the quantum age. Quantum computing operates on the principles of quantum mechanics - a branch of physics that explains the behavior of particles at the infinitesimally small scale. Unlike traditional binary computing systems, these cutting-edge machines can process vast amounts of data and solve complex computational problems at an unprecedented rate.

While it might sound like stuff from a Sci-Fi movie, research and development in quantum computing are making significant strides. Companies such as IBM, Google, and Microsoft are racing towards building scalable, efficient quantum computers. Although still in its infancy, quantum computing holds immense potential for various sectors, including pharmaceutical research, cybersecurity, machine learning, weather forecasting, and financial modeling.

However, the widespread adoption of quantum computing also brings significant challenges - the foremost being cybersecurity. The immense computational power of quantum computers could potentially break contemporary encryption algorithms, making data security a key concern in the future. Hence, parallel to the development of quantum computers, efforts must be undertaken to fortify our cyber-defenses with quantum-resistant encryption.

11.2. The Ubiquitous Internet of Things (IoT)

Our engagement with internet-enabled devices has significantly increased, and this monumental growth has paved the way for the Internet of Things - a network of physical objects embedded with sensors, software, and other technologies to connect and exchange data.

From smart home devices to industrial automation systems, IoT is proving to be a game-changer. Tech giants and startups alike are exploring unprecedented opportunities in this field. Coupled with advancements in AI and machine learning, IoT devices are becoming increasingly autonomous, capable of making decisions without human intervention, thereby enhancing operational efficiency.

However, the impact of IoT is not devoid of challenges. Data privacy, for instance, is a major concern. With billions of devices continuously collecting data, the risk of data breaches escalates. Another concern is the potential job displacement due to automation. Therefore, regulatory guidelines and controls must be in place to address these challenges.

11.3. The Unseen Battles: Cybersecurity

As we become increasingly reliant on digital systems, so does the attractive proposition they make for cybercriminals. Cybersecurity, hence, becomes an absolute necessity. However, emerging technologies are promising hope. Reinforced with machine learning algorithms, cybersecurity systems can now predict, identify, and prevent a wide range of cyber threats.

Artificial Intelligence (AI), for instance, can automate the identification and patching of system vulnerabilities. Blockchain technology provides an unalterable, decentralized ledger system, deterring data tampering and fraud. Biometric authentication systems, on the other hand, can provide a more secure alternative to traditional password systems.

Despite these advancements, the struggle between cybercriminals and security experts continues. Therefore, consistent efforts in research and development and regular software updates are key for the broader implementation of strengthened cybersecurity protocols.

11.4. Decentralizing Trust: Blockchain

Although often associated with cryptocurrencies like Bitcoin, blockchain technology promises far more than a disruptive payment system. At its core, blockchain is a transparent, immutable ledger system that decentralizes trust and can be used to verify transactions in several sectors.

Use cases for blockchain technology are continually expanding, ranging from supply chain management, real estate transactions, healthcare record management, digital voting, and even energy trading. By eliminating the need for intermediaries, blockchain technology increases efficiency, transparency, and security while reducing overall costs.

However, the adoption of blockchain technology also presents challenges including energy consumption, scalability, integration difficulties, and standardization. Therefore, these obstacles need to be addressed to leverage the full potential of blockchain technology.

As we embark on this enticing journey of technological evolution, it's crucial to remember that with every opportunity, there are accompanying risks. While we harness these emerging technologies to make the most of their transformative potential, we must also remain aware and prepared for the new challenges that we'll encounter along the way.

www.ingramcontent.com/pod-product-compliance
Lightning Source LLC
Chambersburg PA
CBHW062313290526

45794CB00006B/2782